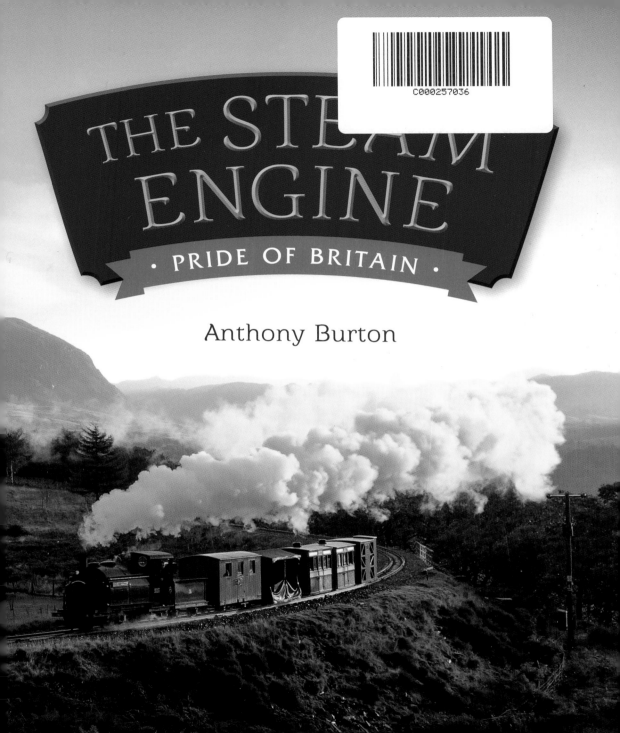

THE STEAM ENGINE

· PRIDE OF BRITAIN ·

Anthony Burton

Important Dates

1st Century AD	Heron of Alexandria uses steam jets to turn a sphere.
1698	Captain Savery demonstrates his engine, *The Miner's Friend*.
1712	Thomas Newcomen's atmospheric engine is installed at Dudley Castle.
1769	James Watt patents his steam engine. Nicolas Cugnot demonstrates a steam tractor.
1783	The first successful experiment takes place with a paddle steamer, *Pyroscaphe*, on the River Saône.
1784	A steam engine is installed to work the machinery of a cotton mill at Papplewick, Nottinghamshire.
1804	The first public demonstration takes places of a steam locomotive running on railed tracks.
1807	The first commercial paddle steamer service begins operations between Albany and New York.
1830	The Liverpool & Manchester Railway opens – the first to use steam locomotives for both freight and passengers.
1831	Sir Goldsworthy Gurney's steam carriage goes into regular service between Cheltenham and Gloucester.
1837	*Sirius* becomes the first vessel to cross the Atlantic using steam power.
1838	Francis Pettit Smith designs a steam launch powered by a propeller.
1843	SS *Great Britain* is launched, the first iron-hulled, propeller-driven steam liner.
c.1850	Steam traction engines are used for ploughing.
1865	The Red Flag Act limits the speed of steam vehicles on the road to 4mph.
1888	A steam turbine is used to generate electricity at Forth Banks Power Station.
1897	The steam turbine vessel *Turbinia* is demonstrated at the Spithead Review.
1904	*City of Truro* becomes the first locomotive to travel above 100mph.
1938	*Mallard* establishes a World Speed Record for steam locomotives of 126mph.
1960	*Evening Star* is the last steam locomotive to be built for British Rail.

Beginnings

The steam engine was to provide the power for almost every aspect of life in the West for over 200 years: for industry; for transport on land and water; for agriculture; and to generate electricity. But its origins stem further back in time. The first record of steam being used to work some form of mechanical device dates back to the 1st century AD. Hero – or Heron – of Alexandria's aeolipile consisted of a sphere containing water that was heated to produce steam. The steam was exhausted through pipes set around the circumference and this made the sphere spin. However, it was more a toy than a practical engine.

The story now leaps forward to the 17th century and experiments not with steam but with air pressure. In 1654 Otto von Geuricke carried out a famous and spectacular experiment: he made two close-fitting hemispheres and, using an air pump, evacuated the air between to create a vacuum. Air pressure of the surrounding atmosphere held the two halves together so strongly that

Hero's steam engine. The fire creates heat under the boiler, and resulting puffs of steam coming out of the nozzles make the globe rotate.

Geuricke's atmospheric pressure experiment was repeated in 1656, using 16 horses in two teams of eight, in his home town of Magdeburg.

30 horses, in two teams of 15, were unable to pull them apart. The idea of using this force in a machine was first developed by Denis Papin. He used a vertical tube, with a little water at the bottom, fitted with a piston. The water was heated and the steam pressure drove the piston up. At the top of the stroke the piston was held by a catch and the cylinder was allowed to cool. The steam condensed back into water and air pressure was allowed to drive the piston back again. It was not a device that had any practical use, but the idea would be developed by others: the combination of condensing steam in a cylinder and the use of air pressure to move a piston.

The Atmospheric Engine

By the end of the 17th century there was a crisis in Britain's mines. To reach the valuable levels of coal or ore, miners needed to go ever deeper, but they were hampered by the lack of suitable pumps to remove the water they encountered. The first successful engine to meet this need was invented by Thomas Savery and patented in 1698 as *The Miner's Friend*. It consisted of a boiler, from which steam was conducted through a pipe to an oval vessel filled with water. When placed near the bottom of the mine, this drove out the water through a second pipe that was then closed off by a valve. The steam, now condensed, created a partial vacuum, and water from the bottom of the mine was sucked up to refill the vessel. There were two vessels that could be used alternately, but there were two problems with Savery's engines: furnaces in gaseous coalmines could be very dangerous and there was a limit to how high the water could be lifted.

The next more efficient engine was invented by Thomas Newcomen of Dartmouth. He was well aware that local tin and copper mines were desperate for an efficient machine to drain their tunnels. His engine used conventional pump rods to lift the water. It was a beam engine, in which the rods were suspended from one end of the overhead beam, automatically pulling that end of the beam down. Just as when you sit on one end of a see-saw, you need someone on the other end to lift you up. So Newcomen now needed a force on the opposite end of his beam. He connected it to a piston in an open-topped cylinder.

Savery's The Miner's Friend: *the water is sucked up from the lowest level, then forced into either of the two vessels marked 'p' and forced out, emerging at the top of the pipe like a fountain.*

Introducing steam below the piston, Newcomen then condensed it by spraying it with water, creating a partial vacuum. Air pressure forced the piston down. The pump rods were lifted and, pressure equalised, fell again. So the beam nodded up and down and the rods rose and fell. The first engine was installed at a colliery at Dudley Castle in 1712. It was a huge success. The engine made 12 strokes a minute, lifting 10 gallons of water over 150ft at each stroke. It had over five times the power of Savery's engine.

The atmospheric engine was to achieve its ultimate form when the pioneer John Smeaton began to apply scientific methods to make improvements. He designed a special boring mill for use at the Carron ironworks in Scotland, and with this and other new machine tools he was able to produce parts to a far higher standard of workmanship. The result was to double the efficiency of the engine.

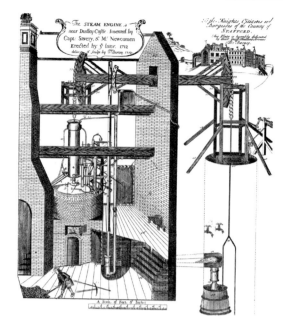

⬆ *The first Newcomen engine was installed at a colliery near Dudley Castle in 1712.*

Success

In the twenty years following the erection of the Dudley engine, over 100 Newcomen engines were installed in Britain.

➡ *The working replica of the Dudley Newcomen engine at the Black Country Museum.*

James Watt

James Watt.

James Watt was born at Greenock in 1736. He trained as an instrument maker and acquired a post as Mathematical Instrument Maker at the University of Glasgow. It was here in 1765 that he was sent a model of a Newcomen engine that refused to work. He recognised that the machine had a fundamental flaw: a huge amount of energy was being wasted alternately heating and cooling the cylinder. It was while out walking on Glasgow Green that the solution came to him. If he condensed steam in a separate vessel, creating a partial vacuum, then steam in the cylinder would, in his own words, 'rush into it, and there be condensed without cooling the cylinder'. The engine could have a separate condenser. But he still had a problem in that heat would inevitably be lost from the open-topped cylinder. If he closed the top he couldn't use air pressure to force the piston down. He realised, however, that he could use steam

➡ *A typical 18th-century Boulton & Watt beam engine: like other early engines it has a wooden beam.*

A close-up of Watt's parallel motion at the Abbey pumping station, Leicester.

pressure instead. He had turned the atmospheric engine into a genuine steam engine.

Watt spent many years trying to develop his idea into a practical machine with the help of a local industrialist Dr John Roebuck, but made little progress. He then received an offer of help from the successful Birmingham industrialist Matthew Boulton. Roebuck offered Boulton a chance to make engines for 'the Midland counties', but Boulton was not interested. He wanted to manufacture engines not for a few counties but 'for the whole world'. In 1773 Roebuck, deep in debt, sold his rights to Boulton and the famous partnership of Boulton & Watt was formed.

Boulton and Watt did not manufacture all the parts for the engines themselves. One of the requirements was to have a close fit between the piston and the cylinder, and for this they relied on the efficient boring machine developed by the ironmaster John Wilkinson, originally intended for boring cannon barrels.

The engines were built on site under the supervision of a Boulton & Watt fitter. The great appeal of the engine was its efficiency, particularly appreciated in the metal mines of Devon and Cornwall, where coal had to be imported at considerable cost. The mines paid an annual premium equivalent to one-third of the saving made by switching from a Newcomen engine. An all-embracing patent taken out by Watt meant that no one else was able to develop steam engines and the company enjoyed a complete monopoly.

It vigorously pursued through the courts anyone who tried to break that monopoly.

Watt's first engine was set to work at Bloomfield Colliery in the Black Country in 1776. Now Watt turned to a new task. This was an age where more and more industries were being mechanised and Watt needed to turn his pumping engine into one capable of turning the wheels of industry itself. The problem was that the piston was attached to the beam by a chain, which could pull down but not push up. He could not simply attach a rigid rod because the end of the beam moved on a circular path. So he developed a system whereby the connecting rod from the piston was attached to a moving parallelogram of rods. He later said that his parallel motion was the invention in which he took the greatest pride. In 1784 a steam engine was used to work the machinery of a cotton mill at Papplewick in Nottinghamshire. It was the start of a major revolution in the manufacturing industry.

Power

On 22 March 1776 the famous biographer of Dr Johnson, James Boswell, described in his diary a visit to the Soho, the Birmingham factory of Matthew Boulton, who greeted him with these memorable words: 'I sell here, sir, what all the world desires to have – POWER.'

New Ideas

lthough Watt's patent should have effectively prevented all experiments with steam engines until it expired in 1800, it did not stop engineers trying out new ideas. The Cornish in particular were accustomed to going their own way and had earlier been responsible for a number of improvements to the older Newcomen engines. One of the first to try a new method was Edward Bull, who developed an inverted engine. Instead of pump rods being attached to the piston via the overhead beam, he turned the steam cylinder upside down over the shaft and attached the rods directly. The case went to court, where Boulton & Watt's man in Cornwall made a neat point. He took a hat and asked what it was; then he turned it upside down and asked, 'What is it now?' It was, of course, still a hat. Bull lost.

The Hornblower family experimented with a very different idea. They passed the exhaust steam from the cylinder to a second cylinder to get extra work from the engine. This idea, known as compounding, would be developed later.

With the end of the prohibitive patent in 1800 the field was open. The first important new developments were down to another Cornish engineer, Richard Trevithick, who had earlier worked with Bull. Watt had been convinced that for safety reasons engines should only use

Rotative engines found many uses: this typical small colliery winding engine can be seen at the Ironbridge Gorge Museum.

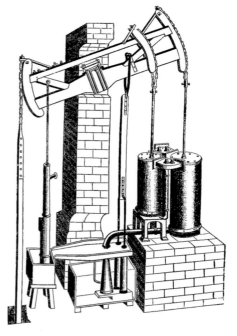

⬆ *A Hornblower compound engine with two cylinders, each with a piston connected to the overhead beam.*

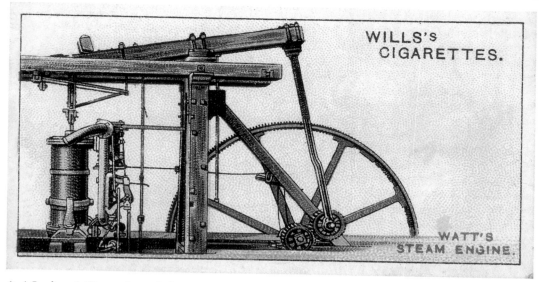

↑ *A Boulton & Watt engine with the sun and planet gear on the right, turning the shaft and large flywheel.*

steam at very low pressure. If more power was needed, then he supplied a bigger engine. Trevithick had another idea. He began to work with high-pressure steam and one of his first tasks was to develop a more efficient boiler, as the ones then in use were little better than overgrown kettles. He designed what became known as the 'Cornish boiler'. It was cylindrical and the hot gases from the firebox were passed through a flue running down the middle of the boiler, heating the surrounding water. Trevithick was able to design engines that were small enough to be moved from place to place and they became known as 'puffers'.

The introduction of high-pressure steam meant that compounding now became a much more successful idea. The exhaust steam from the first cylinder was still at a high enough pressure to make it capable of doing real work. The first engineer to develop the idea was Arthur Woolf. His first attempt at the Meux Brewery in London wasn't a great success, but he eventually produced a two-cylinder engine in 1811 that used only half the fuel of an equivalent Watt engine. The result of all these innovations was to greatly increase the efficiency of pumping engines.

The rotative beam engine saw little change. In order to turn the up-and-down motion of the beam to a circular motion, James Watt had invented a 'sun and planet' mechanism. A connecting rod from the end of the beam ended in a small, toothed wheel that ran round a larger cog, hence the name. It was a clumsy device, soon replaced by a sweep arm and crank. In this form the engine was used for a variety of different tasks. In mines, for example, it was used as a whim or winding engine to move men and materials up and down the shaft.

Throughout the 18th century, virtually all the engines in regular use were beam engines, but in the early 19th century engineers started to develop direct acting engines. There had been numerous attempts to do away with the beam altogether, but the first real breakthrough came from William Symington, who used a horizontal cylinder with the piston attached directly to the crankshaft. One of the most successful of the new devices was the table engine, patented by Henry Maudsley in 1807. This was a vertical engine, with a cross head running between vertical guides. The top of the cross head was attached to the crankshaft. It was to prove very popular, especially in smaller factories.

The Engine Builders

In the early days, when Boulton & Watt were the sole suppliers of engines, they only provided the more complex parts of the mechanism; cylinders were bored by Wilkinson and the mine owners supplied the rest themselves. The company man, in overall charge of the site, supervised the work and had a complete set of printed instructions. Engine constructors at that time were faced with a booklet of some 4,000 words split into twenty different operations. The instructions began with a description of how to build the engine house. This was not just a cover to keep out the rain, but an integral part of the whole structure.

Ruined engine houses are a distinctive part of the Cornish landscape and anyone who looks at one in detail will see that it is very special. One wall is inevitably shorter and far thicker than the other three. This is the 'bob wall' on which the immensely heavy overhead beam or bob pivots. For the biggest engines, bob walls could be up to 6ft thick, and the wall for an engine built for the Tresavean pumping engine used over 500 tons of granite blocks.

One of the problems facing the engine builders was the need to move massive, heavy parts, particularly boilers and beams. One of the biggest manufacturers of engines in the 19th century was Harvey's of Hayle, Cornwall. In 1843 they made the largest steam engine cylinder ever constructed. It was 12ft in diameter, used 25 tons of metal and was to be installed at Cruquis in Holland for draining the Haarlem Mere. Harvey's supplied a great many very large engines, and in order to

A typical Cornish engine house, with the rotative engine under construction, c.1905.

Engines at Dolcoath copper mine, Cornwall in 1831.

The biggest steam cylinder ever cast by Harvey's of Hayle was this 144in monster for the Cruquis engine in Holland.

move them they needed special wagons able to take the load and they kept a stable of 52 draught horses. Moving a large engine was a major operation.

In April 1830 a 15-ton boiler had to be carried up a steep hillside. A team of 20 horses was used to pull the load up the hill, and 100 men were employed to hold on to two long ropes which would act as brakes when needed. Everything was going well until part of the roadway collapsed. Fortunately the wagon broke in two, and only the rear part, together with the boiler, fell over the edge: if it had remained intact, the horses would have plunged down with it. As it was, one of the men became entangled with the rope and fell to his death.

Getting the beam into place was no simple task either. When manoeuvring a large beam weighing many tons, the engineer might arrive to supervise the operation; otherwise it was left to the foreman and his team of perhaps half a dozen or so men to carry out the work. This usually involved either placing stout wooden beams over the top of the engine house walls or erecting shear legs and raising the metalwork with block and tackles. On one famous occasion, simply to surprise the owner, the team installed the beam for an 80in-diameter cylinder overnight by candlelight,

so that when the gentleman arrived the next morning to watch the work get started, they had actually finished.

Engines were robust. For the Boulton & Watt engines, there were instructions for the local blacksmith to make some parts, and the instructions for packing the piston to ensure a good fit with the cylinder sounded quite brutal. Rope had to be beaten flat with a sledgehammer, then put in place and hammered in, after which molten tallow was poured on. When everything was ready, steam was let in and any leaks were closed up by caulking with oakum – the technique used to keep the spaces between boards of a wooden ship watertight.

⬆ *Broken beams with the makers' names at East Pool, Cornwall.*

The Steamboat

Surprisingly, given Britain's dominance in the construction of steam engines, the first successful paddle steamer to take to the waters was launched in France. The inventor was the Marquis Jouffroy d'Abbans, whose vessel *Pyroscaphe* underwent trials on the Saône near Lyons in 1783. The invention was never developed, however, for France was about to be engulfed by political revolution.

The next step was taken by the Scottish engineer William Symington. He installed a small engine in a boat and tried it out on a lake near Dumfries in 1788. One of the first passengers to try this novel form of transport was the poet Robert Burns. Encouraged by his success, Symington was able to persuade the proprietors of the Forth and Clyde Canal to support trials for a steam tug. The *Charlotte Dundas* was fitted with a 22in-diameter cylinder engine. This is the horizontal engine mentioned earlier. It produced a modest 10hp, but was still able to haul two 70-ton barges for 20 miles against a headwind at an average speed of just over 2mph. It was an impressive performance, but the canal authorities were concerned that the churning paddle wheel would destroy the banks, and they abandoned the project.

It was the American Robert Fulton who was to have the honour of establishing the first commercial steamer service, when his vessel *Clermont* began doing regular trips between Albany and New York in 1807. Back in Britain David Napier had taken over his father's Glasgow foundry and begun manufacturing steam engines. In 1812 Henry Bell, a hotelier from Helensburgh, heard about the Fulton experiments and thought he could run a similar service on the Clyde. He asked Napier to design a suitable engine. He built one with a vertical cylinder, with drive transmitted through a pair of side levers to a crankshaft. The vessel, named *Comet*, was a great success, though it seems the engine was often helped out by a sail attached to the tall funnel. It was to be the first of a long line of paddle steamers taking trippers down the Clyde, one of which, the PS *Waverley*, is still delighting passengers, not just on its home river, but all around the British coast.

Below left: The experimental Charlotte Dundas intended for the Forth & Clyde canal, but never used.

Below right: Paddle steamers and sailing ships being built side by side at the Barclay Curle yard on the Clyde in the middle of the 19th century.

Paddle steamer development continued throughout the early part of the 19th century, and in 1836 Isambard Kingdom Brunel, the chief engineer of the Great Western Railway, announced that he saw no reason why he shouldn't extend the line west from Bristol all the way to New York. Many scoffed at the idea, arguing that the ship would consume so much fuel there would be no space left for passengers. But Brunel realised that because the driving force was needed to overcome water resistance, it was proportionate to the area of the hull, not the volume of the ship: the calculations for the hull area were obtained by squaring the numbers; those for volume by cubing them. Doubling the size did not double the fuel consumption: building big was the answer.

The maiden voyage of his paddle steamer *Great Western* in 1837 was delayed and the

⬆ *The triple-expansion engine of the only working sea-going paddle steamer today, PS* Waverley.

tiny steamer *Sirius* beat him to New York. But while *Sirius* had exhausted its fuel supply during the crossing, the *Great Western* had coal to spare. It marked the start of a lucrative trade in ever-bigger transatlantic steamers.

⬇ *Diagram of* Comet's *simple engine.*

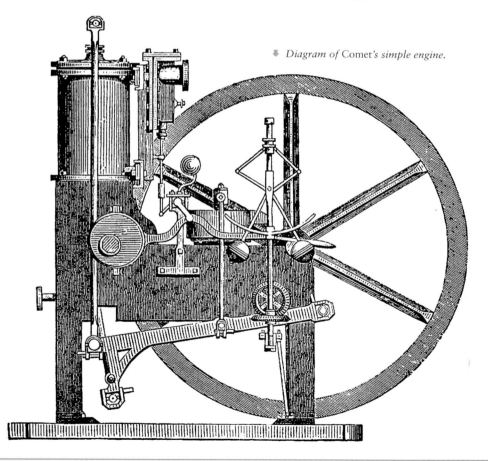

The Steam Carriage

As with the paddle steamer, the first steam carriage to totter off down the road in 1769 was the invention of a Frenchman, Nicolas Cugnot. A former artillery officer, Cugnot's idea was to develop a steam tractor for hauling artillery. It was a clumsy device, with the boiler overhanging the front wheel and, as it had no water feed pump, it had to be stopped every 20 minutes to be refilled – hardly appropriate for a battlefield. It was never developed further. Back in Britain, two men were working quite independently on their own ideas for steam carriages.

William Murdoch was responsible for overseeing the installation of Boulton & Watt engines in Cornwall when he began experimenting with a model steam carriage. Unfortunately for him, his employers got to hear about his work. James Watt was implacably opposed to anyone except himself working on engine development and it was made clear Murdoch had a choice: continue working for the company or continue with the steam carriage. He wisely chose the former course.

At much the same time the Cornish mining engineer Richard Trevithick had begun similar experiments. He started out making a model that was first demonstrated on the tabletop in his kitchen. Due to the Watt patent, he was unable to do any more than that until after 1800 when the patent expired. As we know, he was soon experimenting with high-pressure steam that enabled him to make engines small enough to be portable, being pulled by horses to where they were needed. The next stage was to build a full-sized engine that could move itself.

➡ *William Murdoch's experimental model for a steam carriage.*

↑ *A working replica of Trevithick's 1801 steam carriage.*

The engine was given its first trial on Christmas Eve 1801, when, according to an eyewitness, it went up Camborne Hill in Cornwall 'like a little bird'.

There was a second outing on Boxing Day, when Trevithick and a party of friends set off again. They overturned the little engine but managed to drag it into a building. Unfortunately, as no one remembered to put out the fire in the boiler, the building and the engine were destroyed in the subsequent blaze.

Trevithick soon moved on to a more ambitious project: the London steam carriage. This extraordinary vehicle had a conventional stagecoach body but with a small steam engine slung underneath, providing drive to an enormous pair of 10ft-diameter rear wheels. It was steered by means of a tiller attached to the single front wheel. It was run through the streets of London, but Trevithick's hopes of finding a backer to finance his horseless carriage never materialised. A sea captain who rode in the carriage declared that 'he was more likely to suffer shipwreck on the steam-carriage than on board his vessel'.

Steam transport on the road languished until Goldsworthy Gurney invented his steam drag in 1825, an engine designed as a tractor for use on the roads. The first commercial service using steam power was provided by Walter Hancock's steam omnibus that went into service between Paddington Station and the City of London in 1832. Yet, in spite of this success, the idea was never really developed. The immediate future for steam was not on the road but on railed tracks.

↑ *A working replica of Trevithick's extraordinary London steam carriage.*

The Railway Locomotive

Trevithick may have abandoned the idea of developing a steam carriage for the road, but by 1803 he had a different goal. One of the problems that he never conquered was finding an efficient method of steering his steam carriages: put the engine on rails and the problem disappears. In a letter of August 1802 he made mention of being invited to build a 'locomotive' to run on the tramway at the famous Coalbrookdale ironworks, a track along which trucks were hauled by horses. Not much is known about this engine, but we do know that the same year he was commissioned to build an engine to run on the Penydarren tramway that linked Samuel Homfray's ironworks at Merthyr Tydfil to the Glamorganshire Canal at Abercynon. Homfray's great rival, the ironmaster Richard Crawshay, declared the whole thing impossible and bet 500 guineas that it could not be done, an enormous sum equal to about £20,000 at today's prices. The engine was not considered a railway locomotive as we understand it today: it had to be an all-purpose machine, capable of working machinery at the forge as well as hauling trucks along the track.

⬆ *A replica of Trevithick's Coalbrookdale locomotive is regularly steamed at Ironbridge Gorge Museum.*

⬆ *'Catch-me-who-can' giving passenger rides on a circular track close to the site of the present Euston Station, 1809.*

↑ *The Middleton Colliery Railway, seen by Leeds parish church: the world's first successful commercial railway.*

In February 1804 Trevithick was able to report that the engine had hauled a load of 10 tons, together with '60 or 70' passengers, at a steady 4mph. It was equally successful in its other role: working hammers in the forge. Everything seemed set for a successful future, but one problem soon emerged: the little engine fractured the brittle cast-iron rails.

Trevithick sent another of his engines to a colliery at Gateshead, but the rails there fared no better than they had in South Wales. He was to make one more attempt to persuade the world that the future lay with railway locomotives. He ran the world's first passenger service on a circular track near the present Euston Station in London. The engine, called 'Catch-me-who-can', attracted a lot of visitors but no investors. At this point he abandoned railway ideas.

It was the sharp rise in the price of fodder during the Napoleonic Wars that sparked renewed interest in replacing horses with steam locomotives. The manager of Middleton Colliery near Leeds, John Blenkinsop, decided that he needed locomotives on his tramway that linked the mine to the River Aire. He knew about the problem of breaking rails and decided it could only be solved by increasing the traction for a light engine. Working with local engineer Matthew Murray, he devised a scheme by which a cog on the engine engaged with a toothed rail – the rack and pinion system now only used for mountain railways. It was a success and attracted a lot of attention. Among those who came to see it were Grand Duke Nicholas of Russia, who as Czar would promote Russia's first railway, and a young engineer from a Newcastle colliery, George Stephenson.

Stephenson was not the first to take up the challenge of working on locomotives following the success of the Middleton Colliery line. There were other curious experiments: William and Edward Chapman designed a machine that pulled itself along a chain set between the rails; and William Brunton's engine was supposed to walk along on steam-powered legs. A more conventional approach was adopted by William Hedley of the Wylam Colliery in Northumberland. Two of his engines have survived – *Wylam Dilly* and *Puffing Billy* – both of which had two cylinders set either side of the boiler. Stephenson's first locomotive *Blucher*, which went to work in 1814, was closely modelled on the Middleton engine, but without the rack and pinion system. His fame was assured when he became engineer for the Stockton & Darlington Railway, opened in 1825: the first public railway specifically designed to be worked by steam locomotives.

William Hedley built this locomotive, Wylam Dilly, for Wylam Colliery in 1813. His two sons are posing here for the photograph taken many years later.

Mill Engines

A McNaught compound beam engine at the Bolton Steam Museum.

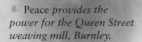

Peace provides the power for the Queen Street weaving mill, Burnley.

By the early 19th century, with the invention of the power loom, all the processes for turning raw material such as cotton and wool into cloth were being taken out of the cottage to be carried out in mills; and water-powered mills were being replaced by steam mills. One of the problems engineers faced was ensuring the engine ran evenly, as it was essential that the machines of the textile industry worked smoothly. The answer was borrowed from the waterwheel. In 1788 Matthew Boulton wrote to his partner James Watt to tell him about a device he had seen – the centrifugal governor – and suggested it might be adapted for steam engines. At first the simple beam engine was adequate, but as mills got ever bigger, so the demand for power increased. One simple answer was to increase the steam pressure by improving boiler design: the Lancashire boiler that came into general use was horizontal, in which the hot gases from the fire passed through a flue, immersed in the water, before passing to the tall mill chimney.

The next stage was to make better use of the steam by building compound engines, in which the exhaust steam from the first cylinder, which was still under some pressure, was fed into a second, low-pressure cylinder built alongside it. The first successful examples were developed by Arthur Woolf, who in 1803 added a high-pressure cylinder to an existing Boulton & Watt engine at a London brewery. It was not an immediate success – he had made the high-pressure cylinder far too small – but he went on to develop the first commercial compound engines.

The next major development came in 1845 when a Glasgow engineer, William McNaught, used a different arrangement. Instead of having the two cylinders side by side, he placed the high-pressure cylinder between the pivot point and the crankshaft. Therefore, when the low-pressure piston was pushing one side of the beam up, the high-pressure cylinder was pulling the opposite side down, and vice versa. One of the advantages of this system was that it was comparatively simple to adapt an existing engine to make it more efficient.

Improvements in engines were made largely because tools for machining with ever-greater accuracy were being introduced, so that, for example, slide valves that controlled the input and exhausting of steam moved smoothly and efficiently. In 1826 a London company, Taylor & Martineau, began developing a simple single-cylinder horizontal engine, instead of the conventional vertical cylinder, for use in factories and mills. Other manufacturers soon followed.

⬆ *Looms and line shafting at Queen Street.*

⬆ *The immense twin tandem compound engine at Trencherfield Mill, Wigan.*

Names

It was customary to give engines a girl's name, which might be anything from that of the owner's wife or daughter to that of a local dignitary. In big compound engines there could be two names. At Trencherfield Mill, for example, the engines to each side of the flywheel are *Rina* and *Helen*. The Queen Street mill engine was installed in 1895, but was renamed in 1918, for obvious reasons, *Peace*.

Over time, steam pressure increased and ever more complex compound engines were introduced, culminating in the twin tandem-compound engines. These consisted of two pairs of cylinders set either side of a huge flywheel. Each pair comprised a high-pressure cylinder set in front of a low-pressure cylinder. One of the most imposing survivors was installed at the former Trencherfield cotton mill in Wigan in 1907. The engine produces an impressive 2,500hp. The flywheel is 26ft in diameter and has 54 grooves to take ropes. These ropes went to turn pulleys on the different overhead shafts situated throughout the mill, from which leather belts conveyed the drive to the different machines.

Rainhill and After

The success of the Stockton & Darlington Railway steam locomotives encouraged promoters to develop far more ambitious schemes. In 1826 Parliament approved an Act for the world's first intercity railway to join Liverpool and Manchester. The company decided that they would be responsible for all the carrying, but there was disagreement about how it should be done.

One faction was in favour of a system of stationary engines, spread out along the line, with trains of carriages and trucks hauled from one to the next by cable. The others wanted steam locomotives. The question to be decided was: could anyone design a locomotive that could cover the whole distance at a reasonable speed, while hauling a train? The answer was to be decided by a public competition. The terms were set. Engines could be up to 6 tons if carried on six wheels and 4.5 tons if on four. They had to be capable of drawing a 20-ton train at 10mph. The contestants were to be put through their paces on a section of the line at Rainhill in 1829.

There were four contestants. The first could be discounted – worked by a horse on a treadmill. The second, *Sans Pareil*, was a sturdy engine, with two vertical cylinders and four connected drive wheels. *Novelty*, the third, was a lightweight vehicle with two cylinders driving a cranked axle. It was described by one of the engineers present as looking like 'an new tea urn', but it was the crowd's favourite, because it was so light and sporty in contrast to the others. Neither was destined to finish the course, however. *Sans Pareil* suffered from a cracked cylinder and retired; *Novelty* proved to have no staying power. They left the way open for the last contestant, *Rocket*, designed by George Stephenson's son Robert.

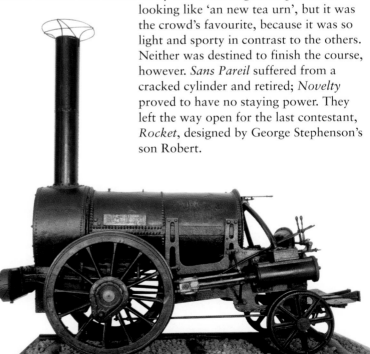

➡ *The remains of* Rocket *in the Science Museum, London.*

➡ *It is an A4 Pacific locomotive that holds the World Speed Record for steam locomotives. This example of the class is named after its designer,* Sir Nigel Gresley, *and is seen here at Rainhill.*

⬅ *The next stage of development from* Rocket: *the replica of Robert Stephenson's* Planet *at Manchester is an obvious forerunner of the modern steam locomotive.*

It was just as well that it won, as it contained all the elements that were used in steam locomotives for the next 100 years. It had a multi-tube boiler: lots of pipes passed from the firebox to carry the heat to the water. The draught to the fire was increased by allowing the exhaust steam to pass up the chimney – an idea first developed by Trevithick – and the cylinders were set at a shallow angle instead of vertically as they had been in all previous engines.

Following the success of *Rocket*, there were to be many improvements and developments to steam locomotives. The first efficient feed pump for replenishing the boiler by forcing in water using steam was invented by the French engineer Henri Gifford. Then in 1841 William Williams, an apprentice draughtsman at Robert Stephenson's Newcastle locomotive works, came up with the idea for a new kind of reversing gear. It not only made reversing easier, but it enabled the driver to exercise control over the steam entering the cylinder.

The valve could be closed before the piston had completed its stroke, allowing the steam to use its expansive power to do extra work.

Over the next half century, the locomotive developed rapidly: steam pressures became ever greater; power was supplied by coupled drive wheels; and the weight of the engine was spread by extra wheels at the front and back. One version, with four front wheels, six drive wheels and two trailing wheels (4–6–2 in conventional notation) was known as an A4 Pacific. It was a locomotive of this type, designed by Sir Nigel Gresley, which established a World Speed Record for steam locomotives that has never been beaten: *Mallard* reached a speed of 126mph near Peterborough in July 1938, while hauling a train of 20 coaches. In 1960 the glory days of the steam locomotive were slowing to a close, and the last steam locomotive to be built in Britain for regular service rolled out of the Swindon works and was given an appropriate name: *Evening Star*.

The Iron Steamer

The first steam-powered vessels to take to the water had all been paddle steamers, and the great majority had conventional wooden hulls. There had been experiments with iron hulls: John Wilkinson had launched an iron barge, appropriately called *The Trial*, on the River Severn in 1787 and Charles Mamby built a larger version in 1822, named *Aaron Mamby* after his father, that became the first iron steamer to go to sea when it crossed the English Channel. But the real impetus for building in iron came when Brunel convincingly demonstrated that for long voyages, vessels had to be built big. There is a limit to the size you can achieve with a wooden vessel but no such restrictions applied to iron hulls. When Brunel decided to follow his *Great Western*, that had made her maiden voyage in 1839, with a larger vessel, the *Great Britain*, he opted for iron. Initially it was still to be a paddle steamer, but a new development in southeast England changed his mind.

A Kentish farmer, Francis Pettit Smith, who enjoyed making model boats, began experimenting with a screw propeller – a device based on the Archimedean screw, which, as its name suggests, had been used for centuries for raising water. Smith's first experiments were carried out on the farm duck pond, but he soon moved to a full-scale trial when he fitted out a launch. He then installed an engine and a screw propeller in a schooner, the *Francis Smith*, that went on trials round the British coast, where it attracted the attention of Brunel.

Work had already begun on designing the engine for Brunel's paddle steamer *Great Britain*, but the engineers had hit a problem: the existing hammers were not up to the job of forging the giant crankshaft the vessel would have needed. The problem was sent to the manufacturer James Nasmyth, who designed a steam hammer. Steam pressure was

➡ *SS* Great Britain *under way: she used both steam and sail power.*

When Nasmyth was set the problem of forging a giant crankshaft, he at once thought of a vertical hammer. He tells the story himself:

Following up this idea, I got out my 'Scheme Book', on the pages of which I generally thought out, with the aid of pen and pencil, such mechanical adaptations as I had conceived in my mind, and was thereby enabled to render them visible. I then rapidly sketched out my Steam Hammer, having it clearly before me in my mind's eye.

Half an hour after receiving the letter outlining what Brunel needed, he had the whole machine sketched out, in just the form in which it would be built.

used to raise the heavy hammerhead vertically and it was then allowed to fall back onto the anvil with tremendous force. It was to prove one of the most valuable steam engines of the age, used in engineering works of all kinds. But once Brunel had decided to change the propeller to a screw design, the need for the giant crankshaft disappeared, and a wholly new type of engine had to be devised to turn the propeller shaft instead. A new age for steamers had begun.

Brunel's steam engine was a somewhat cumbersome beast, with two cylinders forming an inverted 'V' driving the propeller shaft through chains. Steam pressure was very low at just 5 pounds per square inch (psi), so the cylinders had to be enormous: 88in in diameter with a 72in stroke.

As with stationary engines, development of steam-powered vessels was quite rapid, as engineers began to work on developing larger and more efficient marine engines. The chain system was abandoned and later vessels used inverted cylinders, with a simple crank to the shaft. Steam pressure was increased and, by the end of the 19th century, most large steamers had triple-expansion engines, with high-, intermediate- and low-pressure cylinders. The first vessel to use one of these engines was the *Propontis* built by John Elder of Glasgow in 1864. Condensers were also added to later vessels, not so much for their value in increasing the power of the engine, but as a source of pure water for the boiler. Pumping in seawater had a very corrosive effect.

Steam pressure and power are intimately related and a measure of just how far the steam engine had come is the increase in boiler pressure from Brunel's paltry 5psi of 1843 to almost 200psi by the end of the century. The iron steamer had come of age.

⬆ *Nasmyth's steam hammer in action.*

⬆ *The propellor of SS* Great Britain.

Pro Bono Publico

It could be argued that the steam engine did as much to improve public health in the 19th century as any medicine ever did. One result of the Industrial Revolution was the rapid growth of towns and cities, creating a need for good water supplies. New reservoirs were constructed, but in many places the water was pumped up from deep wells, and it was massive steam beam engines that generally carried out this work. The new pumping stations built to house these engines were great sources of public pride and the buildings themselves were often highly decorated – few more so than Papplewick pumping station in Nottinghamshire, opened in 1884. From the outside it has something of the air of a Transylvanian castle, but the interior is a revelation. There are two magnificent rotative beam engines, both by James Watt, that working together could once raise 1,500,000 gallons of water a day. There are stained glass windows with water plants. The main columns are adorned with a tracery of bright metal fishes swimming through metal reeds and, instead of a conventional capital, each column is topped by four ibis. Many of the new Victorian pumping stations, such as that at Kew, used sand filters to ensure that the water that reached customers was free of contamination. Sadly, not everyone in London had access to water of the same quality.

The magnificent Crossness pumping station that has four Watt engines, installed in 1864.

The interior of Papplewick pumping station with a pair of beam engines built by James Watt & Co. in 1884.

In 1854 there was a major cholera outbreak in London's Soho that was to claim 616 lives. At the time it was believed that diseases such as cholera and typhoid were caused by bad air or 'miasma'. It was a local doctor, John Snow, who demonstrated that the disaster was caused not by miasma but by water from the Broad Street pump that was contaminated with sewage. Four years later, even Parliament became aware that something needed to be done. This was the year of 'The Great Stink', when the combination of a hot summer and sewage in the Thames produced such an appalling smell that the windows of Westminster had to be draped in sacking impregnated with chemicals to kill the stench and keep the House of Commons open for business. An Act was quickly passed authorising the construction of a new sewage system for the city.

The man in charge was Sir Joseph Bazalgette, and his plan called for two major sewage systems – one to the north of the river, the other to the south – that would take the sewage far downstream to the east, where it would be collected in holding tanks, ready to be discharged into the river at high tide then washed out to sea. Four pumping stations were built; the one at Crossness, now open to the public, was originally supplied with four engines, each with a 4ft diameter cylinder. In time these proved inadequate and they were later compounded with the addition of both a high-pressure and intermediate-pressure cylinder. Like Papplewick, Crossness is a riot of decorative detail and has often been referred to as a Temple of Steam.

One of the leading suppliers for waterworks was Harvey's of Hayle in Cornwall. They produced a number of immense engines, the largest of which was the 112in cylinder engine for the Battersea pumping station. These are no longer in existence, but two other fine examples, a 100in and a 70in, still survive at Kew, alongside a fine collection of other pumping engines that are regularly steamed.

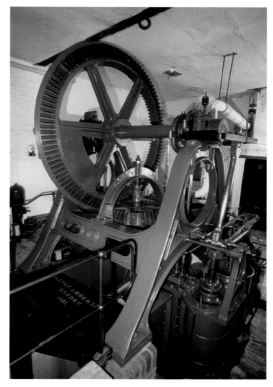

⬆ *This fine 1861 engine by Easton, Amos & Sons was installed at Westonzoyland to pump water from the Somerset Levels.*

Traction Engines

The steam traction engine was a logical development from the portable steam engine that was pulled to where it was needed by a team of horses. It was first developed for the farming community, when it was used to work threshing machines. Robert Ransome established a company in Ipswich to manufacture agricultural equipment and in 1841 developed the very first traction engine. There was still the problem that had bedevilled Trevithick's attempts to promote steam carriages – steering. The solution for the first engine was to have shafts in front and have it steered by horses. Thomas Aveling improved on this system by mounting a single wheel on a low platform in the front of the engine with a seat for a steersman, who moved the wheel by means of a simple tiller. In time a more convenient steering device was developed, with a conventional steering wheel linked to the front wheels. By the second half of the 19th century, specialist threshing teams were working at farms all round the country, bringing the machines to the farm and often taking their own sleeping van with them.

An obvious use for steam power would seem to be ploughing, but there was a problem. You couldn't simply use a traction engine instead of a horse: the heavy engine would do far too much damage to the ground. Various attempts were made to overcome this problem, but it was John Fowler who invented the most successful system. This involved using a pair of engines, each fitted with a cable drum, enabling the cable to be pulled to and fro between the two engines. A plough was attached to the cable, and the ploughshares were

➡ *Ploughing by steam: the ploughshare is being hauled by cable between the two engines.*

⬆ *A traction engine with its threshing set attached.*

reversible, so that the furrows were always turned in the same direction. The two engines moved slowly up either side of the field in unison, with the plough moving first one way then the other.

Traction engines proved ideal for moving heavy loads on the road, but various Red Flag Acts restricted their use. The Acts initially limited speeds to 4mph on the open road, and specified that for long loads a man had to walk in front with a flag.

There was one form of specialist traction engine that combined a whole variety of different uses: the fairground showman's engine. It's first job was to haul the different trucks containing the rides and attractions to the fairground. The arrival of a fair in a community was a big event, and the highly decorated engines with canopies carried on twisted 'barley sugar' columns were very much part of the show. At the fairground,

⬆ *A restored Burrell engine leads a team hauling a heavy load in Dorset.*

the traction engines could mount a small derrick at the front that was then used to erect the different rides. By the end of the 19th century these engines had also taken on a new function: the steam was used to turn a dynamo to produce electricity to light the fair at night.

Showman's engines at Carter's Steam Fair in 2007.

The Steam Turbine

The idea of the steam turbine was not new in the 19th century and the water turbine was already well established. What created the demand for this machine was the development of a new industry – electrical generation.

A dynamo consists of a powerful magnet with a coiled armature of a suitable conducting metal, such as copper, rotating at high speed between the poles. It was not easy to produce high-speed rotation using the typical steam engines of the time, however, as they all needed comparatively complex mechanisms to translate the up and down motion of the piston to rotary motion. In 1884, Charles Parsons, then just 30 years old, began experimenting with adapting the idea of the water turbine to work with steam.

Parsons' first experimental model consisted of a series of turbines in which, rather like the traditional compound engines, each one was adapted to take account of the falling pressure. The turbine comprised of a long central shaft with a ring of blades: the rotor, contained within a casing with another set of blades, the stator. Steam, admitted at one end, passed alternately between the blades of the rotor and stator causing the shaft to turn. Encouraged by the success of his experiments, Parsons was able to take out two patents and in 1888 the very first Parsons turbine with a running speed of 4,800rpm was installed at the Forth Banks Power Station in Newcastle-upon-Tyne. Parsons soon set up his own company, C.A. Parsons and Company, specialising in turbines.

Parsons was well aware that experiments were being carried out by a number of shipbuilders to try and increase the speed of the navy's warships. The speed of a ship is directly related to the speed of the propeller, so he realised that a turbine would be just the thing to achieve a high-speed vessel. In 1894, he formed the Marine Steam Turbine Company and built an experimental vessel, *Turbinia*. She was an amazingly sleek steam yacht, 100ft long but only 9ft beam. The first trials were not a great success, however. The drive had come through a single shaft to the propeller.

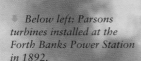

Below left: Parsons turbines installed at the Forth Banks Power Station in 1892.

Below right: The steam yacht Turbinia *at speed in the 1890s.*

THE ISLE OF MAN STEAM PACKET CO. & "MANXMAN" STEAMER SOCIETY PRESENT

"Finished with engines"

YOUR LAST CHANCE TO SAIL ON THE ONLY CLASSIC PASSENGER STEAMER SAILING IN BRITISH WATERS!

Help preserve "Manxman" by supporting this sailing — "Manxman" Steamer Society benefits directly from the revenue earned on this voyage.

FINAL PUBLIC DAY EXCURSION BY T.S.S. **'MANXMAN'** SATURDAY SEPT. 4th

↥ *Poster advertising the very last trip for the only remaining steam turbine passenger ferry* Manxman *in 1982.*

The next model had three shafts turning three propellers. She was unlike anything ever seen before and the performance was astounding. The most famous demonstration occurred at the Spithead Review of 1897 when the little vessel appeared and made a dashing run across the bows of the assembled warships at the unprecedented speed of 34 knots. A popular story has always been that the Admiralty had shown no interest in Parson's novelty, and that the speed run came as a complete surprise. An infuriated admiral sent his fastest vessel to arrest the interloper but couldn't catch her. Sadly, the story is untrue, but it is true that no craft of the time could match *Turbinia* for speed, and a new generation of fast torpedo boats were

built with steam turbines. *Turbinia* is now on display at the Discovery Museum in Newcastle.

The steam launch had a shaft speed of some 20,000rpm, but Parsons realised that if the turbine was to be used for bigger vessels, the speed would need to be reduced, so he introduced geared turbines. His first attempt was a launch that dropped the shaft speed to 1,400rpm. In 1903, the fastest liner afloat was the German *Kaiser Wilhelm II* with two shafts, each driven by massive quadruple-expansion engines. The British response was to build two liners, *Mauretania* and *Lusitania*, with turbines. In 1907 the *Mauretania* achieved the fastest ever time for an Atlantic crossing, a record she was to hold for another 22 years.

Preservation

By the second half of the 20th century, the Steam Age had more or less run its course. The petrol or diesel engine and the electric motor had taken over. But the steam engine has always been loved in a way that few other machines have. There is something very elemental about making something work by lighting a fire and heating water, and the mechanism itself is easy to see and understand. Steam pushes a piston and mechanical links make something work: it is satisfying to see, pleasant to listen to and, for many enthusiasts, smells delightful as well. As a result, as steam engines started to go out of mainstream use, there were people around ready to take them on and put them back to work.

The best-known group is the preserved steam railways, whose story began back in 1946. The Talyllyn Railway was originally built as a narrow-gauge line linking slate quarries in the hills to the port. When work on the quarry stopped in 1946, the owner Sir Henry Haydn Jones kept it going for a while as a passenger line but when he died in 1950 his widow showed little enthusiasm for trying to maintain the rickety track and the collection of care-worn locomotives and carriages. That might have been that if the author L.T.C. Rolt, who had already become involved in canal restoration, had not been walking in the hills. He came across the little line and fell in love with it. Meetings were held, funds were raised and the Talyllyn Railway Preservation Society was formed. Amateurs were about to take over the running of a railway for the first time. It was and still is a huge success, and was soon imitated all over Britain.

An obvious essential for running a steam railway is a supply of steam locomotives, and the preservation movement owes a huge debt of gratitude to the Woodham Brothers and their Barry scrapyard. They bought a lot of

Volunteers on the footplates of a pair of Hunslet industrial locomotives on the West Lancashire Light Railway.

⬇ *Part of the unique collection of engines assembled and restored by the Northern Mill Engine Soicety in Bolton.*

← Traction engines make their way around the main arena at the Great Dorset Steam Fair, where hundreds of period steam traction engines and heavy mechanical equipment from all eras gather for the annual show.

→ The 115-year-old locomotive Derwent *being lifted by a powerful railway crane at Darlington Bank Top Station on its way to to Darlington North Road Locomotive Works to be renovated and repaired in 1961.*

locomotives as they became redundant, but instead of cutting them up for scrap they kept them and made them available for purchase and renovation. Altogether they saved an astonishing total of 213 locomotives, ranging from modest tank engines to main line expresses. Without the Woodham Brothers we would not have the preserved railway network we enjoy today. Equally, the system could not run without a vast number of volunteers, even when lines have permanent staff. The West Somerset Railway, for example, has an army of 900 enthusiasts on their books.

Preserving the big stationary engines is rather more problematical. Many have been incorporated into industrial museums, from colliery winding engines to textile mill engines. Some have been preserved in situ, which has obvious advantages. At Crofton pumping station on the Kennet & Avon Canal, for example, there is a Boulton & Watt engine of 1812 still able to do the job it was installed to do two centuries ago. In other cases, engines have been dismantled and re-erected, as with the magnificent collection put together by the Northern Mill Engine Society and on display in a former mill in Bolton.

The vast majority of traction engines are privately owned and come together on a regular basis at steam fairs held throughout the country, where the general public has the chance to admire them in action.

Several private owners have steam launches, mostly comparatively new, but a few vintage steam vessels exist. There is a wonderful collection of vintage launches at Windermere, the oldest of which dates back to 1850. A few passenger steamers have survived, including the screw steamer *Sir Walter Scott* on Loch Katrine and the paddle steamers *Waverley* and *Kingswear Castle*.

Whether on land or on water, the modern steam enthusiast is blessed with a vast and varied array of engines to marvel at and enjoy.

Places to Visit

There are too many visitor sites to list them all individually, but the following websites provide details of many of the attractions. Steam railways can be found at www.heritagerailways.com and stationary engines are listed, with details, at www.stationarysteamengines.co.uk. The paddle steamer *Waverley* sails from many different ports; her itinerary is at www.waverleyexcursions.co.uk. The following museums and sites can be recommended for their collections of steam engines of various types:

Beamish Open Air Museum
www.beamish.org.uk
Beamish, Durham DH9 0RG

Black Country Living Museum
www.bclm.co.uk
Tipton Road, Dudley DY1 4SQ

Bolton Steam Museum
(Northern Mill Engine Society),
www.nmes.org
Mornington Road, Off Chorley Old Road,
Bolton BL1 4EU

Bressingham Steam & Gardens
www.bressingham.co.uk
Low Road, Bressingham, Diss, Norfolk IP22 2AA

Discovery Museum
www.twmuseums.org.uk/discovery
Blandford Square, Newcastle-upon-Tyne
NE1 4JA

Forncett Industrial Steam Museum
www.forncettsteammuseum.co.uk/Site/Welcome
Forncett St. Mary, Norfolk NR16 1JJ

Kelham Island Museum
www.simt.co.uk/kelham-island-museum
Alma Street, Sheffield S3 8RY

***Kingswear Castle* Paddle Steamer**
www.dartmouthrailriver.co.uk/days-out/
paddle-steamer-kingswear-castle
Kingswear Station Office, The Square,
Kingswear TQ6 0AA

London Museum of Water and Steam
www.waterandsteam.org.uk
Green Dragon Lane, Brentford, London
TW8 0EN

Long Shop Museum
www.longshopmuseum.co.uk
Main Street, Leiston, Suffolk IP16 4ES

Museum of Science and Industry
www.mosi.org.uk
Liverpool Road, Castlefield, Manchester M3 4FP

National Railway Museum
www.nrm.org.uk
Leeman Road, York YO26 4XJ

Riverside Museum
www.glasgowlife.org.uk/museums/riverside
100 Pointhouse Place, Glasgow G3 8RS

Science Museum
www.sciencemuseum.org.uk
Exhibition Road, Kensington, London SW7 2DD

***Sir Walter Scott* steamship**
www.lochkatrine.com
Trossachs Pier, Loch Katrine, By Callander,
Stirling FK17 8HZ

Steam – Museum of the Great Western Railway
www.steam-museum.org.uk
Fire Fly Avenue, Swindon SN2 2EY

Talyllyn Railway Company
www.talyllyn.co.uk
Wharf Station, Tywyn, Gwynedd LL36 9EY, Wales

Thinktank
www.birminghammuseums.org.uk/thinktank
Millenium Point, Curzon Street, Birmingham
B4 7XG

Windermere Steamboat Museum
www.windermerejetty.org
Rayrigg Road, Windermere, Cumbria
LA23 1BN